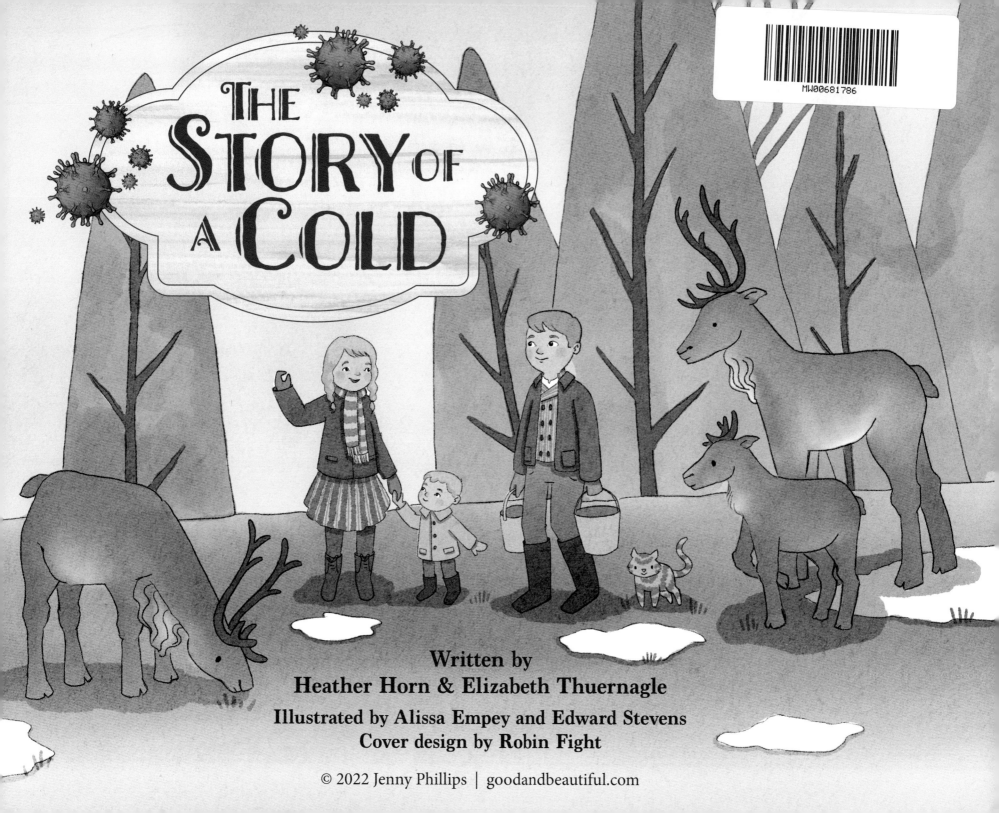

# The Story of a Cold

Written by
**Heather Horn & Elizabeth Thuernagle**

Illustrated by Alissa Empey and Edward Stevens
Cover design by Robin Fight

The sharp north wind bit Ada's nose as she reached for another handful of lichen from the branch above her. Collecting lichen from the forest was something she usually did with her brother Erik, but Mother insisted that he stay inside one more day. Earlier in the month, Erik had gone to visit their cousin Henri for a few days. Well, this visit had not gone as expected! After only two days, Henri had caught a cold due to a virus, and Erik had returned home. A few days later, Erik was stuck in bed with a cold of his own.

## The virus lives in Erik's body.

**Day 1
1:00 p.m.**

A cold is caught when a cold-causing virus enters your body. Viruses (also called germs or microbes) are so small that they can be seen only through the most powerful of microscopes. Though they are small, they can wreak havoc on your body! There are over 200 different kinds of viruses that can give you a cold. That's why colds are so common.

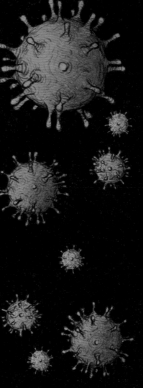

A cold virus can enter your body when you come in contact with an infected person's saliva (spit) or mucus (snot), such as when he or she coughs or sneezes near you. The virus enters your body through your eyes, nose, or mouth. Another way to catch a cold is from touching a contaminated surface. For example, if an infected person sneezes viral particles into his or her hand and then touches a doorknob, the virus can survive on that surface for up to 24 hours. Then, you come along and touch that same doorknob and your face and unknowingly transmit the virus into your body.

3

"Hello," called Ada from the door as she stomped the snow from her boots. She went to Erik's room to check on him and slowly stepped near his bed.

Erik greeted her with many questions. "Did you find much lichen? Did you leave it out where the reindeer can easily find it? The mothers need the extra nourishment for making milk for their calves. *Achoo!*" Erik ended his questioning with a sneeze in Ada's direction.

4

## The virus is transmitted to Ada through Erik's sneeze.

As soon as Ada's brother sneezed in Ada's direction, the virus entered her body through her sinuses. Cold viruses survive in cooler, moist environments, so they thrive in your nasal cavity, throat, and respiratory tract. These body parts are much cooler than the rest of your body because they're in contact with the surrounding air. Once inside Ada's body, the virus latched on to a healthy cell within her sinuses.

5

"Grandmother, this looks wonderful!" Ada complimented. "Reindeer, mashed potatoes, and cranberry sauce—my favorite."

Ada and Erik lived with their parents, their grandparents, and their little brother, Marti, who was just learning to walk. Reindeer herding in Finland had been in Ada's family for several generations, and every family member was needed to help care for and protect the herd.

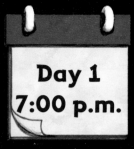

## The virus begins to infect Ada's body.

**Day 1**
**7:00 p.m.**

The virus tricks that healthy cell in Ada's body into allowing it to enter. Once the virus is inside Ada's healthy cell, her cell copies the virus' genetic code, creating many, many more viral particles. These newly copied viral particles are then released into her body, seeking out other healthy cells to trick, and the replication process of the cold virus continues.

At this very early stage, Ada doesn't know she's been infected yet. She is asymptomatic, meaning she doesn't show any of the classic signs of a cold, such as a runny nose. It will take 1-3 days for Ada to start showing symptoms. However, Ada carries the virus, and she can spread it to others.

"We will need to work on that old section of fence on Friday. We need to repair it before next week," Father announced.

"Don't worry, Father. I'm feeling like my old self again, and I'll be up and out the door early to help," assured Erik.

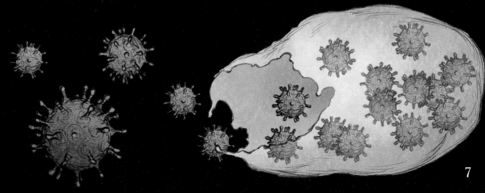

7

Just as Erik had promised, he was already hard at work on the fence by the time Father and Grandfather found him Friday morning. Ada, on the other hand, woke up with a scratchy throat. "They started without me. I'd better get going," she thought as she got dressed.

After lunch Ada and Erik trekked through the forest, gathering more lichen. As Erik reached up for some arboreal lichen, Ada shouted, "Look, Erik! There's a patch of ground lichen hidden under the snow. Let's dig it out to make it easy for the reindeer to find." She ran toward the spot and began digging.

By the end of the day, Erik noticed that Ada's spirited step was now a dragging trudge through the snow as they finished their chores and headed back into the house.

**Ada begins to show signs of illness.**

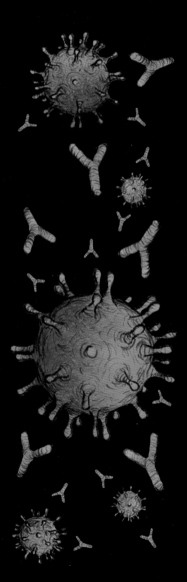

Ada's immune system, now aware that foreign invaders plague her body, begins to create special proteins called antibodies to destroy the virus.

Her scratchy throat is a result of the viral particles infecting her throat and her body responding to them by causing inflammation. Since her body is working extra hard to fight the virus, she feels tired.

"Are you feeling all right, Ada?" Grandmother asked as she greeted Ada and Erik at the door.

"Yeah, I'm just feeling a bit tired and chilled. I am sure it is from all that digging we did today," she replied.

"If I didn't know better, I'd say you caught my cold. Mine started with a scratchy throat. I thought I was fine, but then I got really tired, and then I had the chills," recalled Erik.

"Oh no, I woke up with a scratchy throat, but I didn't say anything," Ada said.

"It sounds to me like Erik's right. We need to begin treatment right away," said Grandmother.

Ada's immune system continues to work hard to fight off the virus, but it's a process that takes time and energy. Meanwhile, Ada feels sick. One way her body fights the virus is by increasing her body temperature through burning extra energy, resulting in her feeling a bit chilled. Remember that the virus is still attacking the healthy cells in her sinuses, throat, and other parts of her respiratory system.

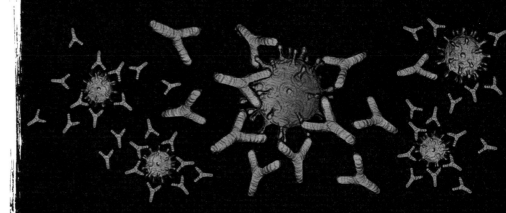

"Are you saying that I need to go to the doctor?" asked Ada.

"No. There are many things we can do here at home to take care of you and to help you fight the virus," Grandmother said.

"I have a virus? I thought I had a cold!" replied Ada.

With a calming voice, Grandmother assured her, "Come, Ada, let's get you tucked into bed. I'll make you some soup to warm you up and tell you all about it."

Ada snuggled under her covers and closed her eyes. She heard a gentle knock at her door. "Come in, Grandmother."

"Here, I brought you some *Lohikeitto,*" Grandmother said as she handed Ada the warm bowl. "It was my grandmother's salmon soup recipe. She made it for me when I was a sick girl. It will give you nourishment to keep your strength up. You will need it. A cold can last for a week or more."

"Wait a minute, Grandmother. I thought you said I have a virus," Ada questioned.

Grandmother gave Ada a smile and explained, "A cold is a type of virus. There are germs all around us, in the air, in our food and water, and on the things we touch. They are very tiny and can be seen only under a microscope. There are two main groups of germs: bacteria and viruses. A cold is a virus. A virus cannot live outside a host. It needs energy from a host to grow and reproduce, so if there is a virus on something or in the air, it does all it can to enter into a host."

## Ada's body shows additional signs of the virus.

Ada's body continues to host a battle between her immune system and the high number of viral invaders. Her immune system valiantly works hard to seek out and destroy the virus, while the viral particles keep replicating and attacking healthy cells.

To find the viral particles, Ada's immune system sends out an army of white blood cells that identifies and engulfs viruses. How do her white blood cells do this? The little antibodies latch on to proteins found on the outer casing of the viral particles. The antibodies attached to the viral particles become beacons, signals for other white blood cells to come and destroy.

While this is going on inside her body, resting allows Ada to conserve energy so that her  body can use the energy to fight the virus. The warm fluids from Grandmother's soup help soothe Ada, keeping her hydrated and her mucus loose.

13

"What is a host?"

"A person or animal that the virus invades. In your case, you," Grandmother answered. "Since we know Erik had a cold earlier, I am sure you picked up the virus from him."

Ada remembered his greeting earlier in the week. "Erik sneezed in my direction the other day. Do you think that may be where I got it?"

"That sounds like a good possibility. A sneeze is one of the natural ways our bodies kick germs out. This was good for Erik, but not good for you. His sneeze forcefully pushed the germs out of him and into the air. As you breathed the air into your lungs, you became the new host."

"Tell me more, Grandmother."

"Tomorrow. The best thing you can do right now is get some good sleep. While you sleep, your body will be hard at work fighting off the virus." Grandmother turned out the lights and said good night.

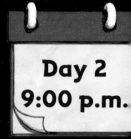

**Ada falls asleep, but her body is still fighting the virus.**

**Day 2
9:00 p.m.**

As Ada sleeps, her immune system is strengthened, her body is repaired, and energy is conserved. The viral particles are being hunted down, tagged, and destroyed by Ada's white blood cells.

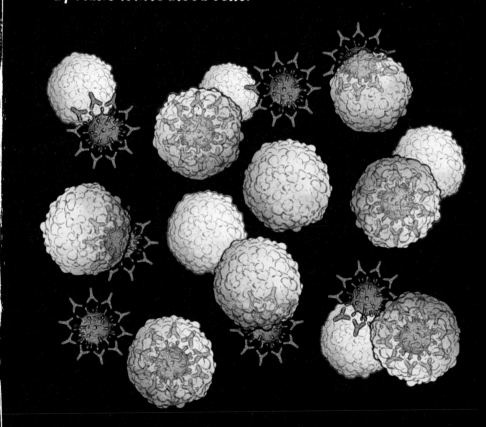

"Good morning, Ada. How are you feeling?" Father asked as he checked on Ada.

"I feel worse than last night. I slept pretty well, but I still feel so tired. Sorry I overslept. I'll be out to help with chores soon," Ada answered with a nasally voice.

"Oh no you won't, sweet girl! You are to stay in this bed all day and rest. Here, you sound like you will be needing these today," Father said as he handed Ada a box of tissues.

**Ada is more congested and feels rotten.**

While fighting a cold virus, it's common for your body to produce a lot of mucus. It may feel annoying to you as the mucus clogs up and drains out of your sinuses. Sometimes you might hear the mucus rattle during a cough, and mucus can sometimes interfere with your voice box, causing your voice to sound funny. Mucus also drains to the back of your throat, causing your throat to feel raw and sore. The swelling reaction of your body's immune system also contributes to a sore throat.

So why does your body produce so much extra mucus during a cold? Having this cold tells your body to release a chemical called histamine, ramping up the mucus production. This clear substance helps trap the virus and is also full of antibodies. The sticky quality of mucus is the result of a protein called mucin, which sticks together to create a tough shield against viruses.

"My nose is all stuffed up, and my face feels like a reindeer stepped on it," Ada sniffled.

"You are getting congested. The virus is multiplying in your sinuses. Your immune system is already hard at work."

"My immune system?" Ada wondered.

"When God designed our bodies, he gave us an immune system. Its job is to fight infections and prevent you from getting the same infection again in the future. Your cells are being attacked by the virus, which makes your blood vessels swell. This is why it feels like you have been stepped on by a reindeer. It is actually a good sign. Your body is sending a large number of white blood cells to your sinuses. They are working hard to destroy the virus. I'll send Grandmother in to take good care of you today. She knows many ways to help your body heal itself from a cold virus. You will be in good hands."

## Ada's congestion grows thicker.

Even when you're not sick, your body naturally produces mucus all the time to prevent parts of your body, such as the nasal passage, from drying out. On average, you produce more than a liter per day of total body mucus! But when you're sick, your body produces more.

Ada may have also noticed that her mucus was beginning to turn yellow or green. This is a sign that her body is doing its job properly. This coloration comes from flushed-out dead white blood cells that have successfully trapped the viral particles.

While the mucus is indeed helpful, it causes the classic symptoms of a cold, including clogged sinuses, which contribute to congestion, pressure, and pain (such as sinus headaches).

Sneezing, coughing, and blowing the nose are also common with a cold. This happens because the viral particles within the mucus are being flushed out. Mucus is essential in keeping viruses at bay.

19

Ada closed her eyes for what she thought was a few minutes, but when Grandmother came in, her clock told her otherwise. She had fallen asleep for two hours!

With a sniffle and a sneeze, Ada sat up to greet Grandmother.

Grandmother set down the tray she was holding and handed Ada a tissue. "I have brought you some chicken soup, crackers, fruit, and water. A healthy meal with nutrients like vitamin C and zinc helps keep your immune system strong. I also want you to drink lots of water today. Taking in lots of fluids like soup, water, and juice will help your body flush out the virus. Keep that box of tissues handy—you will need it."

She was right; Ada's wastepaper basket was soon filled with used tissues. Ada's congested nose turned into a flowing faucet. She spent the afternoon coughing, sneezing, and blowing her nose. Her body was kicking out the virus.

## Ada begins sneezing and sniffling.

**Day 3
1:15 p.m.**

The viral particles are still attacking Ada's respiratory system, and her body is responding through mucus production and other common symptoms.

By the second or third day, the virus is typically at its highest count, so symptoms are usually the worst. The cold symptoms mentioned are actually caused by your body's response to the virus.

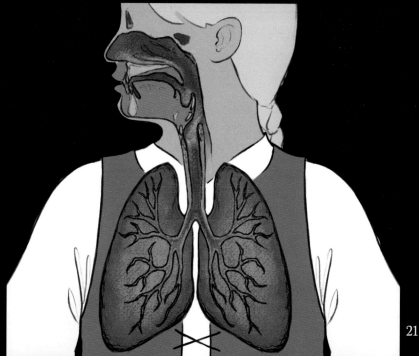

That night Ada had a hard time sleeping. Her throat was burning, and her barking cough did not help. Her runny nose made it very hard to breathe. Tossing and turning, she kicked off her blanket and called for her mother.

"I'll draw you a bath. It will help you relax and help bring your fever down. Having a fever is not fun, but it is another way your body is naturally fighting off the virus. Viruses multiply rapidly at certain temperatures, but if they are made too hot or too cold, they die off. Drinking lots of fluid helps prevent you from becoming dehydrated as you sweat away the fever. It also helps keep your throat and sinuses moist to make it easier to sneeze, blow, and cough out the virus. After your bath, I'll bring you some warm lemon tea with honey. It will soothe your throat."

## Ada's temperature rises.

**Day 3
11:00 p.m.**

When viruses enter the body, the body reacts by releasing chemicals called pyrogens, which make body temperatures go up. At a higher temperature, your body's chemical reactions tend to slow. The healthy cells that have been overtaken by a virus will slow in their ability to copy the viral code, reducing the number of viral particles being copied.

Fever is less likely in a common cold than with the flu, but in an attempt to try to kill off the virus, the body may spike its temperature.

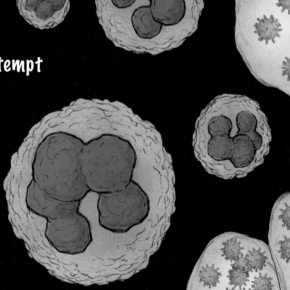

23

Over the next few days, Grandmother took good care of Ada, refilling her water, bringing her food, and answering her questions. Grandmother also spent lots of time with Ada, telling her stories from her childhood. But Ada's favorite times were when Grandmother played her *kantele*, a Finnish harp, while singing her to sleep.

Ada's body continues to fight the virus as she rests.

Fluids keep the mucus loose and hydrate the body. Eating healthy foods that have vitamins and minerals supports and strengthens the immune system.

Blood vessels often enlarge during a cold to allow more white blood cells into the affected area, resulting in puffiness and redness.

Resting allows the body to have all the energy it needs to completely kill off the virus. If people don't rest enough and start to be more active because they feel better, they may start to feel sick again because they still haven't fully recovered.

25

As the days passed, Ada slowly felt better. Her fever broke, she had more energy, and her coughing and sneezing let up. She was finally well enough to spend more time with her family.

"Good morning, everyone! I am feeling so much better, thanks to Grandmother," Ada announced to everyone as she came into the kitchen.

26

"And to your strong immune system!" Grandmother added. "We all need to help our bodies stay strong with a healthy breakfast. Now go wash your hands with warm, soapy water to wash away the germs so they do not get passed to the rest of us."

Ada's fever breaks, and she starts feeling better.

Now Ada's immune system has the antibodies for that virus, so if she accidentally passes it on to Grandmother, Grandmother can't pass it back to her because she's already immune to that specific virus. This is also why Ada's brother Erik couldn't catch her cold again, because he had already had that cold. However, remember there are over 200 different strains of the common cold, and currently, she's immune to only one strain.

After breakfast Ada played with Marti while Mother and Grandmother cleaned. They said it was very important to disinfect the surfaces and air in the house to help eliminate any leftover germs. As Ada reached for Marti's favorite blocks, she felt a sneeze coming on. She learned her lesson from Erik. She quickly turned her head away from Marti and into her elbow. *"Achoo!"* She caught it. This sneeze was not going to pass any germs on to Marti if she could help it. She went to the bathroom to wash up.

## The house is disinfected.

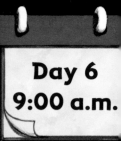

Sanitizing or washing your hands with soap breaks down the protective barrier around viruses and bacteria, killing them. Wiping down surfaces also helps you avoid introducing a virus through touch.

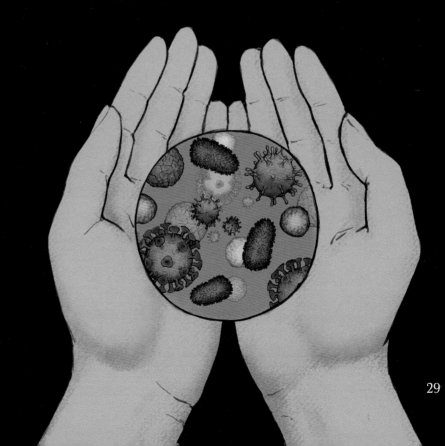

29

When she returned to the living room, she saw Marti pointing out the large picture window. In the distance she saw Erik waving. Next to him was a mother reindeer and her new calf. Ada could hardly contain her excitement. Spring was here, and many more calves would soon be born.

"Marti, let's get bundled up and go out for some fresh air. We will get a closer look at our new friend. I learned a lot about cold viruses, and the fresh air and exercise will help keep us healthy. Come on; I'll tell you all about it."

## Ada gets some fresh air.

Day 6
10:00 a.m.

In general, going outside for some fresh air makes you less likely to breathe in recirculated air from within a home or building. However, this won't guarantee you won't get sick. Exercise helps boost your immune system.

31